The Shock Of Change
that understanding the
human condition brings

Jeremy Griffith

Watch the video of this presentation at
www.HumanCondition.com/SOC

OR

Scan code to view

The Shock Of Change that understanding the human condition brings
by Jeremy Griffith

Published in 2022, by WTM Publishing and Communications Pty Ltd
(ACN 103 136 778) (www.wtmpublishing.com).

All enquiries to:

WORLD TRANSFORMATION MOVEMENT®
Email: info@worldtransformation.com
Website: www.humancondition.com or www.worldtransformation.com

The World Transformation Movement (WTM) is a global not-for-profit movement
represented by WTM charities and centres around the world.

ISBN 978-1-74129-077-6
CIP – Biology, Philosophy, Psychology, Health

Filming and editing by James Press.
Cover image: *Humanity's Journey from Ignorance to Enlightenment* by
Jeremy Griffith

The drawings by Jeremy Griffith, copyright © Fedmex Pty Ltd (ACN 096 099 286)
1991-2022. Computer graphic of exiting cave by Jeremy Griffith, Marcus Rowell
and Genevieve Salter, copyright © Fedmex Pty Ltd (ACN 096 099 286) 2009.

Commendations for Griffith's treatise

From Thought Leaders

'[**Prof. Stephen Hawking**] is most interested in your impressive proposal.'
● 'In all of written history there are only 2 or 3 people who've been able to think on this scale about the human condition.' **Prof. Anthony Barnett**, zoologist
● '*FREEDOM* is the book that saves the world…cometh the hour, cometh the man.' **Prof. Harry Prosen**, former Pres. Canadian Psychiatric Assn. ● 'I am stunned and honored to have lived to see the coming of "Darwin II".' **Prof. Stuart Hurlbert**, esteemed ecologist ● 'Living without this understanding is like living back in the stone age, that's how massive the change it brings is!' **Prof. Karen Riley**, clinical pharmacist ● 'Frankly, I am blown away by the ground-breaking significance of this work.' **Prof. Patricia Glazebrook**, philosopher ● 'I've no doubt a fascinating television series could be made based upon this.' **Sir David Attenborough** ● '*FREEDOM* is the necessary breakthrough in the critical issue of needing to understand ourselves.' **Prof. David J. Chivers**, former Pres. Primate Society of Britain ● 'Whack! Wham! I was converted by Griffith's erudite explanation for our behaviour.' **Macushla O'Loan**, *Executive Women's Report* ● 'This is indeed impressive.' **Dr Roger Lewin**, preeminent science writer ● 'I have recommended Griffith's work for his razor-sharp biological clarifications.' **Prof. Scott Churchill**, psychologist ● 'An original and inspiring understanding of us.' **Prof. Charles Birch**, zoologist ● 'The insights are fascinating and pertinent and must be disseminated.' **Dr George Schaller**, preeminent biologist ● 'Very impressive, particularly liked the primatology section.' **Prof. Stephen Oppenheimer**, geneticist, author *Out of Eden* ● 'I consider the book to be the work of a prophet.' **Dr Ron Strahan**, former dir. Sydney Taronga Zoo ● 'The scholarly value [of Griffith's synthesis] is comparable to several of the most celebrated publications in biology.' **Prof. Walter Hartwig**, anthropologist ● 'I believe you're on to getting answers to much that has bewildered humans.' **Dr Ian Player**, famous Sth. Afr. conservationist ● 'A superb book, a forward view of a world of humans no longer in naked competition.' **Prof. John Morton**, zoologist ● 'This might bring about a paradigm shift in the self-image of humanity.' **Prof. Mihaly Csikszentmihalyi**, psychologist ● 'As a therapist this is a simply brilliant explanation.' **Jayson Firmager**, founder of *Holistic Therapist Magazine* ● 'The questions you raise stagger me into silence; most admirable.' **Ian Frazier**, author *Great Plains* bestseller ● 'The WTM is an island of sanity in a sea of madness.' **Tim Macartney-Snape**, world-leading mountaineer & twice Order of Australia recipient

Commendations From The General Public

'Griffith should be given Nobel prizes for peace, biology, medicine; actually every Nobel prize there is!' ● 'He nailed it, nailed the whole thing, just like the world going from FLAT to ROUND, BOOM the WHOLE WORLD CHANGES, no joke.' ● '*FREEDOM* will be the most influential, world-changing book in history, and time will now be delineated as BG, before Griffith, or AG, after Griffith.' ● 'I'm speechless – this is bigger than natural selection & the theory of relativity!' ● 'I really think this man will become recognized as the best thinker this world's ever seen, and don't we need him right now!' ● 'Griffith has decoded the human species, we FINALLY know what's going on & the suffering stops!' ● 'The world can't deny this for much longer, let the light in, save the human race!' ● 'This is the most exciting moment in my life. *THE Interview* tore my hat off & let my brain fly into the sky!' ● '*THE Interview* should be globally broadcast daily. The healing explanation humans so sorely need.' ● 'In a world that's lost its way there's no greater breakthrough, water to a world dying of thirst.' ● 'Dawn has come at Midnight! A brilliant exposition, we could be on the cusp of regaining Paradise!' ● 'This man has broken the great silence, defeated our denial, got the truth up, woken us from a great trance.' ● 'Beware the 'deaf effect; your mind will initially resist the issue of our corrupted condition and so find it hard to take in or hear what's being said, but if you're patient you'll find the redeeming explanation of our condition pure relief.' ● 'John Lennon pleaded "just give me some truth", well this site finally gives us *all* the truth!' ● '*FREEDOM* is the most profound book since the Bible, now with the redeeming truth about us humans.' ● '*Death by Dogma* is brilliant clarification.' ● 'We were given a computer brain, but no program for it; but Aha, Griffith has found it, made sense of our lives!' ● 'This just goes deeper & deeper in explaining us, like dawn devouring darkness, amazing!' ● 'Agree, this is not another deluded, pseudo idealistic, PC, 'woke', false start to a better world, but the human-condition-resolved real solution.' ● 'Freedom indeed! What we have here is the second coming of innocence who exposes us but sets us free!' ● 'As prophesised, King Arthur has returned to save us (mentioned in par.1036 *Freedom*)' ● 'We all need to go back to school & learn this truthful explanation of life.' ● 'Join in our jubilation, your magic reunites, all men become brothers, all good all bad, be embraced millions! This kiss [of understanding] for the whole world' – From Beethoven's 9th (par.1049 *Freedom*)

Contents

With the real problem of the
human condition finally solved we
can now ACTUALLY fix the world!

Background

Jeremy Griffith is an Australian biologist who has dedicated his life to bringing fully accountable, biological understanding to the dilemma of the human condition—the underlying issue in all human life of our species' extraordinary capacity for what has been called 'good' and 'evil'.

Jeremy has published over ten books on the human condition, including:

— *Beyond The Human Condition* (1991), his widely acclaimed second book;

— *A Species In Denial* (2003), an Australasian bestseller;

— *FREEDOM: The End Of The Human Condition* (2016), his definitive treatise;

— *THE Interview* (2020), the transcript of acclaimed British actor and broadcaster Craig Conway's world-changing and world-saving interview with Jeremy about his book *FREEDOM*;

— *Death by Dogma: The biological reason why the Left is leading us to extinction, and the solution* (2021), which presents the biological reason why Critical Theory threatens to destroy the human race;

— *The Great Guilt that causes the Deaf Effect* (2022), which describes how lifting the great burden of guilt from the human race initially causes a 'Deaf Effect' difficulty taking in or 'hearing' what's being presented;

— *Therapy For The Human Condition* (2023), which is about the therapy that is desperately needed to rehabilitate the human race from our psychologically upset state or condition, elaborating on what is presented in *FREEDOM*;

— *Our Meaning* (2023), which explains how being able to know and fulfil the great objective and meaning of human existence finally ends human suffering;

— *The G*reat Transformation: How understanding the human condition actually transforms the human race (2023), which gives a concise description of how the psychological rehabilitation of humans occurs, and how everyone's life can immediately be transformed;

— *AI, Aliens & Conspiracies*: The Truthful Analysis (2023), which provides Jeremy's thoughts on the much discussed question of the danger of Artificial Intelligence (AI), and on the possibility of alien life visiting Earth, and also his explanation for the epidemic of conspiracy theories; and

— *Sermon On The Beach* (2024), which is Jeremy's elaborated transcript of his inspired description of how the human race now leaves the horror of the human condition forever!

This booklet, ***The Shock Of Change that understanding the human condition brings*** is a transcript of a presentation Jeremy gave in 2022, which can be viewed at www.humancondition.com/SOC. It's about managing the change from ignorance to enlightenment of our human condition.

Jeremy's work has attracted the support of such eminent scientists as the former President of the Canadian Psychiatric Association Professor Harry Prosen, the esteemed American ecologist Professor Stuart Hurlbert, Australia's Templeton Prize-winning biologist Professor Charles Birch, the former President of the Primate Society of Great Britain Professor David Chivers, Nobel Prize-winning physicist Professor Stephen Hawking, as well as other distinguished thinkers such as the pre-eminent philosopher Sir Laurens van der Post.

Jeremy is the founder and patron of the World Transformation Movement (WTM)—see www.humancondition.com.

The Shock Of Change
that understanding the human condition brings

[1] Welcome to this July 2022 presentation about managing the shock of change that inevitably occurs with the arrival now of the all-wonderful, all-meaningful and all-exciting human-condition-resolved new world for you, and for the whole human race. (You might like to print out the transcript of this talk to help you follow what will be explained.)

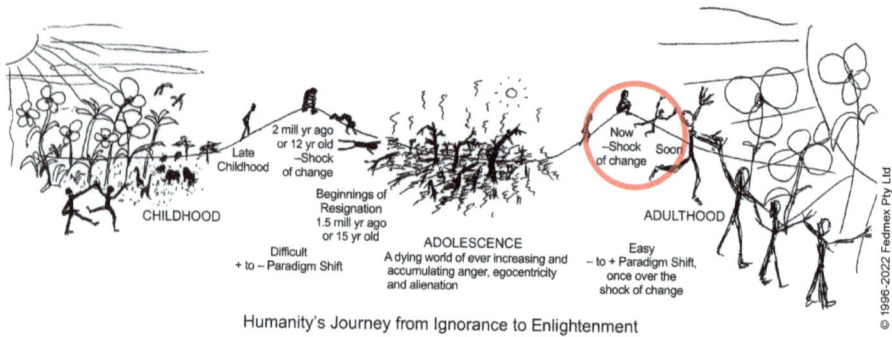

2 mill yr ago
or 12 yr old
−Shock
of change

Late
Childhood

CHILDHOOD

Difficult
+ to − Paradigm Shift

Beginnings of
Resignation
1.5 mill yr ago
or 15 yr old

ADOLESCENCE
A dying world of ever increasing and
accumulating anger, egocentricity
and alienation

Now
−Shock
of change   Soon

ADULTHOOD

Easy
− to + Paradigm Shift,
once over the
shock of change

© 1996-2022 Fedmex Pty Ltd

Humanity's Journey from Ignorance to Enlightenment

[2] This 'shock of change' stage is highlighted with a red circle in my drawing of *Humanity's Journey from Ignorance to Enlightenment*. You can see that humanity has progressed from an innocent but ignorant childhood; then, when we became conscious some 2 million years ago, through an upset angry, egocentric and alienated human-condition-afflicted adolescence—adolescence being the insecure stage in which we search for our identity or meaning, particularly for understanding the reason why we corrupted our original state of innocence. With the finding of that understanding, humanity enters an enlightened adulthood in which we are secure in our understanding of ourselves. What the red circle highlights is the initial shock of change that the human race is now going through following the arrival of this dreamed-of ability to finally understand ourselves. As I said, this talk is about managing that particular paradigm-shifting shock of change.

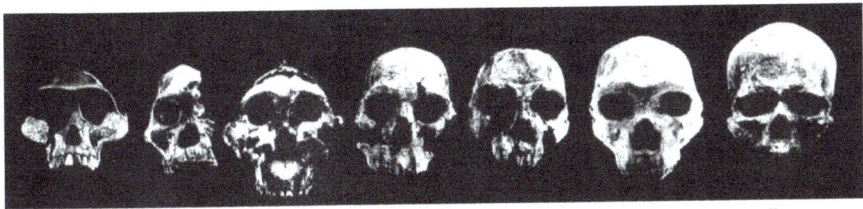

Australopithecus afarensis	*Australopithecus africanus*	*Australopithecus boisei*	*Homo habilis*	*Homo erectus*	*Homo sapiens*	*Homo sapiens sapiens*
Fossil evidence from 3.9 to 3 million years ago	3.3 to 2.1 m y a	2.3 to 1.2 m y a	2.4 to 1.4 m y a	1.9 to 0.1 m y a	0.5 to 0.1 m y a	0.2 m y a to now
Brain Volume 400 cc average	450 cc	530 cc	650 cc	900-1100 cc	1350 cc	1400 cc
Early Happy Childman ➡	Middle Demonstrative Childman ➡	Late Naughty Childman ➡	Distressed Adolescentman ➡	Adventurous Adolescentman ➡	Angry Adolescentman ➡	Pseudo idealistic and Hollow Adolescentman

(Human Condition Fully Emerges Here)

Humanity's stages of maturation, see chapter 8 of *FREEDOM*. (Note our big brain with its large **'association** [thinking] **cortex'** appeared some 2 mya.)

[3] Before getting into this talk about the shock of change that understanding the human condition brings, I want to show you this picture of the fossilised skulls of our ancestors under which I have included descriptions of the psychological stages of maturation that each ancestor was going through. These stages are all explained in chapter 8 of my book *FREEDOM*. This sequence also indicates that our big brain with its large **'association** [thinking, conscious] **cortex'** appeared some 2 million years ago, which corresponds to the top of the first hill in my *Humanity's Journey* picture. The red zig-zag is where the upset human condition began with the emergence of consciousness and its conflict with our instincts. Interestingly, the pronounced neotenised, child-like, dome-forehead, wide-eyed, snub-nose features of us humans today, *Homo sapiens sapiens*, also indicates how much we've been chasing the image of innocence for sexual destruction in the latter part of our 2-million-year journey through upset, a tragic development that is explained in paragraph 787 of *FREEDOM*. To now address the issue of the immense shock of change that the arrival of the dreamed-of, human-race-liberating understanding of the human condition can't help but cause.

- - - - - - - - - - - - - - - - - -

[4]Firstly, I need to emphasise that during our 2-million-year adolescence stage (which we are just coming out of) when we couldn't explain our corrupted condition, we had no choice other than to live in denial of it. Until we found understanding of our corrupted condition, denial of it saved us from unbearable condemnation.

[5]As I point out in all my writings, without the redeeming understanding of our species' upset, corrupted, 'fallen' human condition—and of our own upset, corrupted, 'fallen' condition as a result of it—it has been impossible for virtually anyone to confront and be honest about the human condition. We couldn't face and be honest about the corruption of our original all-loving and all-sensitive instinctive self or soul while we couldn't explain why we were seemingly so stupid, irresponsible and at fault for destroying such a beautiful existence. We couldn't face the truth of our corrupted condition while we were unable to defend ourselves against the implication that behaving in such a non-loving, divisive, competitive and aggressive, psychologically upset way meant we were bad, worthless, even evil, beings.

[6]This situation left us collectively and individually with no choice but to keep adding more and more layers of protective denial to cope with the horror of our ever-increasing soul-corrupted and soul-corrupting condition—because any time we tried to go back and confront and be honest about our corruption only reconnected us with the unbearably depressing implication that we were bad, worthless, soul-destroying, evil monsters. So this is where the human race has been trapped, having no choice but to accept that until the human race's great, heroic upsetting search for knowledge finally led to the finding of the redeeming, relieving and healing, good reason for why we corrupted our soul, almost everyone had no choice but to 'resign' themselves to having to hide in Plato's dark cave of denial of our soul-corrupted and soul-corrupting human condition. (The process of Resignation is explained in Freedom Essay 30, and Plato's cave allegory is described in Video/F. Essay 11.)

[7]Certainly we could gain some artificial reinforcement and relief for our immensely distressed, guilt-ridden, insecure situation by winning as much power, fame, fortune and glory as we could. And certainly, through taking up support of a religion where we deferred to the soundness and truthful words of a prophet, we could to some degree counter the dishonest denial of our corrupted condition. And we could certainly find some superficial, makes-you-feel-good relief from being so corrupted by supporting a pseudo idealistic movement like the politically correct, 'woke' movement, or the environmental green, climatism movement. And we could certainly try to superficially therapise our corrupted condition by calming our distressed mind through such practices as meditation, or by taking mind-altering drugs, or by undertaking a course in positive thinking, or even a course in mind-disconnecting non-thinking. But such artificial reinforcement of our corrupted condition, or partial honesty about being corrupted, or escapist relief from being corrupted, or superficial therapy for our corrupted condition, still left virtually everyone firmly stuck in Plato's dark cave of denial, unable to confront and be honest about the unbearably depressing issue of our soul-corrupted and soul-corrupting, seemingly-human-race-destroying human condition.

Computer graphic by James Press © 2018 Fedmex Pty Ltd

[8] Without the redeeming biological explanation for our corrupted condition, denial of it was all that has been possible for virtually everyone. So yes, we could find any number of ways to achieve some superficial and artificial relief from the shame and insecurity caused by our species' 2 million year corrupted condition, but, without the redeeming explanation for it, virtually everyone has been trapped having to live out their lives as a desperately insecure, almost totally dishonest and deluded person! The truth is, human life has been a life of excruciating torture.

[9] Sure, we had to try to hide the fact that we have been living such a horrible existence by putting on a brave face and maintaining an outwardly happy and positive appearance—because, again, if we were honest about our horrifically soul-corrupted and soul-corrupting condition it only made living in that state unbearably, suicidally depressing. Denial has been our stock in trade, our only way of coping. Being stuck in Plato's dark cave of denial and pretence and delusion has been the life virtually all humans have heroically had to endure for some 2 million years!

"Putting on a brave face" (actors in the Broadway musical *On the Town*)

[10] As F. Essay 30 on Resignation explains, during adolescence almost everyone had no choice but to resign themselves to blocking out the truth of their and almost everyone else's horrifically soul-corrupted and soul-corrupting condition, and instead pretend that they were living a happy, secure, guilt-free, sound, well-adjusted, everything-is-fine, I'm-a-good-person life. But every child who came into that world of resigned adults knew it was an almost totally dishonest and deluded, seemingly completely mad, furiously angry, extremely soul-destroyed, utterly artificial and superficial world. These pre-resigned lyrics from the young American heavy metal band With Life In Mind reveal the truth about the *real* nature of our world: 'It scares me to death to think of what I have become…I feel so lost in this world', 'Our innocence is lost', 'I scream to the sky but my words get lost along the way. I can't express all the hate that's led me here and all the filth that swallows us whole. I don't want to be part of all this insanity. Famine and death. Pestilence and war. A world shrouded in darkness…Fear is driven into our minds everywhere we look', 'Trying so hard for a life with such little purpose…Lost in oblivion', 'Everything you've been told has been a lie…We've all been asleep since the beginning of time. Why are we so scared to use our minds?', 'Keep pretending; soon enough things will crumble to the ground…If they could only see the truth they would coil in disgust', 'How do we save ourselves from this misery…So desperate for the answers…We're straining on the last bit of hope we have left. No one hears our cries. And no one sees us screaming', 'This is the end' (*Grievances* album, 2010).

[11] This is a similar truthful description of the mad-with-pain-and-anger, utterly dishonest and deluded resigned world that we have been living in, this time from Christ who had one of the most honest, denial-free, sound minds in recorded history (see my demystification of Christ in F. Essay 39): 'O bitterness of the fire that blazes in the bodies of men and in their marrow, kindling in them night and day, and burning the limbs of men and making their minds become drunk and their souls become deranged…Woe to you, captives, for you are bound

in caverns [so this is another Plato-like reference to being imprisoned in a cave]! **You laugh! In mad laughter you rejoice! You neither realize your perdition, nor do you reflect on your circumstances, nor have you understood that you dwell in darkness and death! On the contrary, you are drunk with the fire and full of bitterness. Your mind is deranged on account of the burning that is in you, and sweet to you are the poison and the blows of your enemies! And the darkness rose for you like the light, for you surrendered your freedom for servitude! You darkened your hearts and surrendered your thoughts to folly, and you filled your thoughts with the smoke of the fire that is in you! And your light has hidden in the cloud [of darkness] and the garment that is put upon you, you [pursued] [with deceit]. And you were seized by the hope that does not exist. And whom is it you have believed? Do you not know that you all dwell among those who [lie] [... and you boast] as though [you had hope]. You baptized your souls in the water of darkness! You walked by your own whims!'** This amazingly honest quote comes from the *Book of Thomas*, which is a recording of a conversation Christ had with one of his disciples called Thomas. (As an aside, I think it's interesting that earlier in this conversation, when Christ says to Thomas **'it has been said that you are my twin and true companion'**, I think Christ is referring to Thomas being one of the rare unresigned people like himself, and such a person would have been a **'true companion'** in Christ's lonely task of defying the cave-dwelling, dishonest world of denial that we live in.)

[12] The great Scottish psychiatrist R.D. Laing was equally honest about our corrupted, soul-denying-and-repressing, split-from-our-true-self-or-soul, alienated condition when he wrote that **'Our alienation goes to the roots. The realization of this is the essential springboard for any serious reflection on any aspect of present inter-human life...We are born into a world where alienation awaits us. We are potentially men, but are in an alienated state** [p.12 of 156] **...the *ordinary* person is a shrivelled, desiccated fragment of what a person can be. As adults, we have forgotten most of our childhood, not only its contents but its flavour; as men of the world, we hardly know of the existence of the inner world** [p.22] **...The condition of alienation, of being asleep, of being unconscious, of being out of one's mind, is the condition**

of the normal man [p.24] ...between *us* and It [our soul] there is a veil which is more like fifty feet of solid concrete. *Deus absconditus* [God has absconded]. Or [more precisely] we have absconded [from God/the integrative ideal state] [p.118] ...The outer divorced from any illumination from the inner is in a state of darkness. We are in an age of darkness. The state of outer darkness is a state of sin—i.e. alienation or estrangement from the inner light [p.116] ... We are all murderers and prostitutes...We are bemused and crazed creatures, strangers to our true selves, to one another' [pp.11-12] (*The Politics of Experience* and *The Bird of Paradise*, 1967). 'We are dead, but think we are alive. We are asleep, but think we are awake. We are dreaming, but take our dreams to be reality. We are the halt, lame, blind, deaf, the sick. But we are doubly unconscious. We are *so* ill that we no longer feel ill, as in many terminal illnesses. We are mad, but have no insight [into the fact of our madness]' (*Self and Others*, 1961, p.38 of 192). 'We are so out of touch with this realm [where the issue of our corrupted human condition lies] that many people can now argue seriously that it does not exist' (*The Politics of Experience* and *The Bird of Paradise*, p.105).

[13] So that is the true reality of our immensely insecure, soul-blocking-out, alienated human situation or condition!

- - - - - - - - - - - - - - - -

[14] What I now want to describe is how being able to explain and understand our corrupted condition ends the need for denial of it, and how a whole new world opens up for the human race where we are free of the agony of the human condition.

[15] Finding the redeeming, guilt-lifting, good reason for our horrifically corrupted, angry, egocentric and alienated human condition completely changes the situation we have been in. NOW THAT WE FINALLY HAVE THE FULLY ACCOUNTABLE AND THUS VERIFIABLE AND THUS TRUSTABLE FIRST-PRINCIPLE-BASED, BIOLOGICAL, SCIENTIFIC EXPLANATION OF THE GOOD REASON FOR WHY WE CORRUPTED OUR ORIGINAL ALL-LOVING AND ALL-SENSITIVE INSTINCTIVE SELF OR SOUL, OUR WHOLE 2-MILLION-YEAR TORTUROUS CONDITION ENDS!

[16] With understanding of the human condition at last found we are no longer trapped in Plato's cold, dark cave of denial and delusion, and are free to come out into the warm healing sunshine of the compassionate full truth of the redeeming and dignifying understanding of ourselves. Our insecurity about being corrupted ends. We can finally understand that we are all good and not bad. And what this means is that ALL OUR OLD ARTIFICIAL AND SUPERFICIAL DISHONEST AND DELUDED WAYS OF COPING ARE NO LONGER NEEDED. We are FREE to leave all those fake and in-truth horrid practices behind as finished with. We can finally transition from living in a dreadful state of denial and delusion to living in a wonderfully TRANSFORMED STATE of freedom from all the dishonesty and delusion that made human life so fraudulent, mad, destructive and chaotic. In fact, all the in-truth extreme shallowness, emptiness and loneliness of that cold and dark dishonest and deluded, immensely destructive existence is able to be replaced by so much relief, happiness and healthiness now that we will find it almost too exciting to bear! We go from living an existence tortured by the horror of the human condition, to a life free of that horror.

William Blake's *Cringing in Terror* (c.1794–96) left, and *Albion Arose* (c.1794–96) right

[17]Basically, we humans now come back to life from having had to endure living in an effectively dead state. As the descriptions from With Life In Mind, Christ and R.D. Laing made clear, 2 million years of accumulating upset *has* left us in a numb, seared effectively dead state—but we are finally free now from having to live trapped in Plato's deathly dark, deep, cold cave of denial and delusion. The whole human race wakes up from what has in truth been a terrible, terrible nightmare. WE ARE FREE TO ALL COME BACK TO LIFE NOW; the great TRANSFORMATION of the whole human race is on!

[18]In chapter 9 of my book *FREEDOM* I describe how understanding the human condition obsoletes the deluded and artificial ways of proving we are good and not bad (such as by winning as much power, fame, fortune and glory as we can, or by supporting a pseudo idealistic cause like the environmental or the PC, 'woke' movements that make us feel we are a good person), and allows everyone to take up a fabulously relieving and healing TRANSFORMED WAY OF LIVING where you live in support of the redeeming understanding of the human condition. While chapter 9 of *FREEDOM* gives an overall description of how this completely wonderful transformation takes place, Video/F. Essay 33 provides a specific description of how to let go

and leave the old insecure, have-to-prove-you-are-good-and-not-bad ways of living and adopt the new, human-condition-free Transformed Way Of Living where you can live a fabulously meaningful and joyous life in support of the redeeming understanding of our corrupted condition. [Note, since publishing this book in 2022, Jeremy Griffith and others have presented a concise description of how everyone's life can immediately be transformed in the booklet/video *The Great Transformation* and in the booklet/video *Sermon On The Beach*, both of which appear on the Transformation page of the WTM's website.]

[19] What immediately needs to be emphasised here is that with humanity's immensely heroic but at the same time immensely upsetting search for the redeeming explanation of our corrupted condition finally over, not only CAN everyone leave the old, now obsoleted, ways of sustaining their sense of self-worth and instead adopt the new Transformed Way Of Living, everyone actually HAS TO leave that old way of living and instead live in support of understanding of the human condition. What makes everyone HAVE TO make this transition is not the dictates from some outside authoritarian regime, or the imposition of a dogma-based, pseudo-idealistic culture, but the imposition of the unassailable logic that the old selfish artificial ways of living are destroying the world and making human life completely unbearable, and since those ways of living are now no longer needed there is NO JUSTIFICATION AT ALL for continuing to live that way. ADOPTING THE TRANSFORMED WAY OF LIVING IS WHAT EVERYONE CAN AND THEREFORE HAS TO DO NOW TO SAVE HUMANKIND.

[20] I should clarify, as I do in paragraph 1231 of *FREEDOM* and in F. Essay 33, that the focus in leaving the old world of artificial reinforcement isn't on giving up your possessions or walking the streets in sackcloth in self-denial and servitude. We're talking about a change of mindset that can have an effect on your priorities, which can affect your choice of possessions and so forth, but the focus isn't on self-deprivation.

[21] I also want to point out that in my book *Death by Dogma*, in paragraphs 94-96, I describe how we humans have always known that one day we had to mature from living in an insecure, defensive

competitive and aggressive state to living in a secure cooperative and loving state, but being unable to confront the human condition and by so doing find the understanding that actually makes that transformation possible, out of sheer desperation it was decided to just 'fake it to make it', which is what the Critical-Theory-based, PC, 'woke', 'Great Reset' movement was trying to do. While such fakeness was never going to work—as *Death by Dogma* makes clear—it does reveal how intuitively anticipated and needed this great transformation is.

[22] So yes, taking up the Transformed Way Of Living is all-important—that is absolutely true and fundamental—but the focus of this presentation is on what everyone will initially particularly want to know, and that is HOW ARE WE TO COPE WITH THE INITIAL SHOCK OF HAVING OUR 2-MILLION-YEAR CORRUPTED CONDITION SUDDENLY EXPOSED? How are we to manage the initial inevitable shock that the human race has to go through on the way to becoming transformed?

[23] As I depicted in my drawing of the stages humanity goes through in its progression from ignorance to enlightenment, it is inevitable that when understanding of the human condition finally arrives after 2 million years of living without it and having had no choice but to live in denial of the whole issue of our corrupted condition, the resigned world is going to be in shock at what has happened. Having become deeply attached to and accustomed to living in denial of our corrupted condition, having the truth about that state suddenly revealed *cannot help* but come as a great shock.

- - - - - - - - - - - - - - - -

[24] So I now want to present the focus of this talk, which is the 'shock of change' stage that the human race now goes through.

[25] This shock of change progresses through a number of predictable stages, which I've illustrated with some drawings.

[26] It makes sense that initially everyone will be stunned for a little while, staggered by what has happened and in a state of overwhelmed

astonishment, needing time to take in, absorb and adapt to what has occurred. The arrival of the understanding of ourselves that the whole human race has so heroically fought so long for isn't initially met with celebration, excitement and relief but by a great silent, NONPLUSSED PARALYSIS. In his famous 1970 book *Future Shock*, Alvin Toffler was anticipating the initial bewildering effects of the arrival of understanding of the human condition when he wrote of **'the shattering stress and disorientation that we induce in individuals by subjecting them to too much change in too short a time'** (p.4 of 505).

[27] Of course, as I make very clear in my booklet *The Great Guilt*, what the minds of resigned adults quickly do to cope with this shock of having their corrupted condition exposed is determinedly apply the denial they have been employing up to this point in their life to hold that truth at bay; basically their mind doesn't tolerate admission of it. The result of this defensive blocking out of description and analysis of the human condition in the minds of resigned adults is that they typically find it difficult to take in or absorb or 'hear' what it is being presented. Their mind doesn't let what's being said penetrate their conscious awareness. It suffers from what we refer to as the 'DEAF EFFECT'. And this protective 'deafness' can be extreme; as one resigned adult admitted, **'When I first read this material all I saw were a lot of black marks on white paper'**!

The initial 'Deaf Effect' reaction to description of the human condition

[28] Since this denial is happening unwittingly, subconsciously—because the surface conscious mind obviously doesn't want to recognise and admit that it is being defensive—the surface conscious mind's natural response is to blame the presentation for its inability to 'hear' it. Typical responses include saying that the presentation is **'impenetrably dense'**, and **'too repetitive of vague points'**, and so badly written it **'desperately needs editing'**. In fact, the Deaf Effect is quite often so great that in frustration we are asked to **'please send me an executive summary because I can't make any sense of what you're presenting'**!

[29] However, this Deaf Effect can be overcome by PERSEVERING with reading about and listening to descriptions of the redeeming, reconciling and healing explanation and how it makes sense of every aspect of human existence. With perseverance, the resigned mind gradually realises that what is being presented is not condemning but compassionate, and that it is therefore now safe to learn about and finally understand the whole 2-million-year immensely corrupting but also incredibly heroic journey the human race has been on to find understanding of its corrupted condition. The following is a good example of someone patiently re-reading and re-listening to the explanation and overcoming the Deaf Effect. May Gibbs, the founder of the WTM Phoenix Centre in Arizona, America said, **'I tried for months to get through the reading material but my thought at the time was that it was super tedious and boring.'** Saying it was **'super tedious and boring'** is a typical Deaf Effect response. The subject of the human condition has historically been so unbearably confronting and depressing that most people's minds just don't want to engage with it, and so defensively and dismissively, in effect, say, 'I'm not interested in what you're talking about, it's just super tedious and boring, meaningless rubbish as far as I'm concerned!' Significantly, when May persevered listening to and reading the information and eventually got through the Deaf Effect she said, **'I find it laughable now...** [how deaf I was because] **here I am today...obsessing over the Freedom Essays, videos and Facebook Group posts...It has brought such peace to my life and**

I have a burning desire to get it to whoever will listen.' And this is another comment about how perseverance overcomes the Deaf Effect. It's from a post by Melinda Greenacre on the WTM's Facebook Group: '**Hello, I'm a first time poster. Something weird is happening – I thought I'd share. I first came across this book [***FREEDOM***] in 2017 when I downloaded the free version. I couldn't get past the first few pages. Then during lockdown in 2020, I bought the paperback version for myself and my Dad, but still couldn't get through the first couple of pages. It just didn't make any sense, I felt that the sentences were too long and rambling. Reading it made me tired, my mind would wander away every time, so I put it down again. Recently** [in 2023] **after seeing lots of WTM content pop up on social media I started playing the main pinned interviews** [in the Facebook Group] **in the morning as I was getting ready for work. A few tiny bits of information clicked into place, nothing mind blowing, but definitely a few "ah ha" moments. Today, I've been sitting at the bay waiting for my husband to finish a charity bike ride. I bought a coffee and thought it would be a good location to try reading the book again. In just over an hour, I've gone through 50 pages. I don't even remember why I found it so difficult, all of a sudden it's making sense and I'm enjoying it. It's like I'm reading a different book!'** (21 Aug. 2023). (See *The Great Guilt* video/essay and Video/F. Essay 11 for an in-depth description of the Deaf Effect.)

Persevering enables you to overcome the 'Deaf Effect' and access understanding of every aspect of human existence

Drawing by Jeremy Griffith © 1996-2023 Fedmex Pty Ltd

[30] What will make all the difference to the initial huge problem of so many people suffering from the Deaf Effect and being unable to 'hear' and appreciate these human-race-saving understandings of the human condition is a critical mass of people appearing who have overcome the Deaf Effect and realised, and are telling others, that the human condition *has* finally been solved. People's willingness to persevere with listening to and reading about the human condition rapidly increases once there are enough people saying that the redeeming and healing explanation of the human condition has finally been found and that it is now both safe and necessary for everyone to learn about the human condition. So overcoming the huge problem of the Deaf Effect block and gaining wide public acceptance and support for this absolutely critically important breakthrough depends on building a critical mass of support.

[31] Also of immense significance to humanity completing its journey to enlightenment is that once a person is able to overcome the Deaf Effect and can access and thus appreciate the explanation of the human condition, they are then in a position to take up the TRANSFORMED WAY OF LIVING where they selflessly live in support of the understandings of the human condition rather than selfishly maintaining their old, now obsoleted artificial, power and glory and pseudo idealistic ways of sustaining their sense of self-worth.

The human-race-saving and all-exciting Transformed State

Drawing by Jeremy Griffith © 1996-2023 Fedmex Pty Ltd

[32]Naturally, this transformation from the old habituated, denial-based way of living doesn't happen immediately. The first response, once a person is able to get through the Deaf Effect, is to keep reading about and listening to the explanation of the human condition and by so doing be able to make sense of every aspect of human life. Having lived without any truthful, accountable understanding of human existence, it is *extremely* interesting, relieving and healing to finally have such insight. So this GUZZLING OF THE TRUTH about human life normally goes on in earnest in this initial stage. When resigned adults have more or less given up trying to find understanding of the nature and purpose of human existence (because in their resigned state of living in denial of the human condition, all resigned adults subconsciously knew that all the mechanistic scientific explanations of human life are based on the same dishonest denial that they are practising), suddenly there are so many accountable and clearly truthful understandings to digest. (You can read more about mechanistic science's dishonest denial of the human condition in chapter 2:4 of *FREEDOM*.)

[33]However, even though all these explanations about human existence are based on the compassionate, redeeming understanding of the good reason for why we humans became corrupted, these insights are not only going to be immensely interesting, they are necessarily also going to be exposing and confronting of all that corruption.

[34]Ideally, a person would be able to make sense of and heal all the soul-damaging hurts they experienced growing up in a human-condition-stricken world at the same speed as they are able to understand the human condition, but the reality is that understanding of our corrupted condition occurs much faster than we can connect with and dismantle all the layers of repressed psychological pain within us. In fact, it is going to take a number of generations for the understandings we now have of the human condition to heal all the psychosis in us humans. Knowing to the point of being psychologically secure follows some time after understanding, and the degree we are still insecure is the degree we can feel confronted and exposed

by the truth of our corrupted condition. The truth day, honesty day, transparency day, exposure day, revelation day, come-clean day that the arrival of understanding of the human condition represents is actually the long feared so-called 'JUDGMENT DAY' anticipated in religious texts. And even though 'judgment day' is actually a time of compassionate understanding, not a time of condemnation—as an anonymous Turkish poet once said, judgment day is **'Not the day of judgment but the day of understanding'** (*National Geographic*, Nov. 1987)—it is nevertheless a time when the extent of our corrupted condition is revealed.

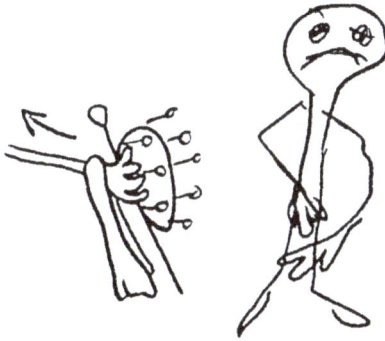

'Truth Day' or 'Honesty Day' or 'Revelation Day'
or 'Exposure Day' or 'Judgment Day'

[35]Naturally then, the more upset a person is as a result of their particular encounter with humanity's heroic but upsetting search for knowledge, the more upset there is to be exposed and therefore the more confronting these understandings of the human condition will initially likely be. Even though upset has now been explained as an immensely heroic condition, and that everyone is equally good and worthwhile, this shock of the sudden exposure of our upset condition can initially cause almost everyone, but the more upset in particular, to want to defensively retreat back into the habituated safety of Plato's dark cave of denial. There will be a temptation to want to deny and even determinedly ATTACK ANALYSIS OF THE HUMAN CONDITION. Indeed, we in the WTM have had to endure decades of these malicious

and often ruthless attacks upon us and our project of delivering the human-race-saving understanding of the human condition—attacks you can read about in the *Persecution of the WTM* essay.

Attacking the truth about the human condition

[36] However, since this understanding of the human condition is what is desperately needed to save the human race from terminal alienation and extinction, retreating back into denial of all the fully accountable, truthful understandings is not a viable option for anyone—so <u>how then is the upset human race, the more upset in particular, to cope with the exposure? The obvious answer is that everyone needs to assess how much of the truthful explanation of the human condition they can cope with, and avoid confronting more of it than their degree of soundness of self can cope with.</u>

[37] What this means is that the process of hearing and reading about the truthful explanation of all aspects of human life that is now available has to be tempered by how much truth a person can cope with—'guzzling' can be dangerous. <u>The reality is that everyone only needs to investigate these explanations of the human condition sufficiently to verify that they are the truthful explanations that the human race has been desperately in search of and then avoid continuing to investigate and study them beyond what their security of self can cope with.</u> On reaching that point what each person

needs to do is change their focus from studying the understandings to supporting them. Everyone should VERIFY THE TRUTH THEN SUPPORT IT AT A COMFORTABLE DISTANCE FROM IT.

Supporting the truth without overly confronting
it and by so doing saving the human race

[38] And living in support of these human-race-saving under-standings is a completely satisfying and meaningful existence, so there is no particular negative or penalty or down-side in not overly studying them. In fact, the Transformed Way Of Living where we focus on living in support of these world-saving understandings is so exciting and fulfilling that, as I said earlier, it is almost more than our body can bear.

The human-race-saving and all-exciting Transformed State

[39] So the Transformed Way Of Living or State serves two purposes: it is how we leave the old selfish, now-human-race-destroying, artificial ways of sustaining our sense of self-worth, and it is how we avoid overly studying explanation of the human condition. And I should point out that, as I mention in paragraph 1188 of *FREEDOM*, in the old evasive world to enter university you had to pass entrance exams that assessed your IQ (intelligence quotient) and resulting ability to study complex subjects like higher mathematics and physics; well, in the human-condition-resolved new world it similarly makes sense that the exceptionally sound, those with a high SQ (soul or soundness quotient), are the ones most suited to studying the human condition. And there is absolutely no shame in that demarcation. Again, a fundamental insight that understanding of the human condition gives us is the equal goodness and worthiness of all humans. We humans have been involved in a great and necessary battle, so we have inevitably all been variously knocked around in that great battle, but we are all heroes, with those who do not have a high SQ arguably the greatest heroes, because they must have been involved in the thick of the battle during their infancy and childhood.

[40] In chapter 9:7 of *FREEDOM* I explain in even more detail how exposure of our corrupted condition can be managed. And it should be pointed out that the human race would not have so determinedly fought to find understanding of our corrupted human condition if it didn't believe we would be able to cope with the exposure of it when it was found. So many of our songs that anticipate our freedom from the human condition are songs of rejoicement, and they wouldn't be so joyful if we thought that freedom couldn't be coped with. (See F. Essay 45 about songs that joyously anticipate our freedom from the human condition.)

[41] Another problem that develops as people digest these understandings of the human condition is what we refer to as the 'MEXICAN STANDOFF'. This is where a person knows and accepts that the explanation of the human condition is true, but struggles to let go of their old, now obsoleted, and now extremely selfish, destructive and

dangerous ways of sustaining their sense of self-worth and instead adopt the new Transformed Way Of Living. People can feel stuck between not wanting to retreat back into denial but not yet being appreciative enough of the immense importance and exhilarating benefits of taking up the Transformed Way Of Living. (Note, FAQ 1.23 provides a powerful description of the problem of the Mexican Standoff. And in Video/F. Essay 33, which I referred to earlier, I explain how to overcome the Mexican Standoff by letting go of the now obsoleted but often extremely strongly habituated attachment resigned adults have to their power, fame, fortune and glory 'trophy room' or 'ego castle' of pre-human-condition-solved 'old world' 'wins' and 'successes', or of attachment to pseudo idealistic 'virtue signalling' strategies, and instead taking up the now infinitely more satisfying and world-savingly important Transformed Way Of Living.)

J.Griffith © 2019 Fedmex Pty Ltd

The 'Mexican Standoff' situation where you don't want to deny the truth but also don't want to let go your now obsoleted ways of sustaining your sense of self-worth

[42] As with overcoming the problem of the Deaf Effect, the more people there are who appreciate that the Transformed Way Of Living is the only way to solve the world's problems, and also how relieving and exciting a way of living it is, the more encouraged and thus easier it will be for others to take up the Transformed Way Of Living. In fact, as I point out in paragraph 1207 of *FREEDOM* where I talk about the importance and power of the Transformed

Way Of Living, "before long those still living in the old embattled, have-to-prove-your-worth way of living will feel like they have been caught wearing last year's fashions!"

[43] What now needs to be explained is that while the fabulously meaningful and joyous human-race-saving new 'fashion' of the Transformed State is having to go through this inevitable but actually pointless initial procrastination, Mexican Standoff stage, there are a number of devious manoeuvres and avoidance strategies that occur, which we in the WTM are very familiar with.

[44] There is what we call 'POCKETING THE WIN' where people's attachment to their egocentric way of living in which they sustain their sense of self-worth by building an 'ego castle' of power, fame, fortune and glory wins can be so great that they avoid taking up the Transformed Way Of Living and instead focus on finding ways to selfishly benefit from understanding the human condition.

[45] This includes the blatant exploitation of the understanding where people use the insights they have gained into human behaviour to be an even more successful winner of power, fame, fortune and glory. It is obviously a big advantage in business or competition to have insight into human behaviour, but exploiting that to make yourself successful and financially rich in the 'old world' is an obscene misuse of this world-saving information.

The 'pocket the win' strategy where you 'rip off'
the insights to win even more power and glory

J.Griffith © 1996-2018 Fedmex Pty Ltd

[46] As well as almost literally 'pocketing the win' by ripping off the understandings to make yourself an 'old world' legend, there are other 'pocketing the win' responses that are more focused on finding ways to escape feeling cornered by the logic that says you have to let go your artificial ways of sustaining your sense of self-worth and instead take up the Transformed Way Of Living. We call such escapes from taking up the Transformed Way Of Living 'SIDE DOORING'. Feeling cornered by having to let go your now obsoleted but habituated castle-building way of validating yourself does tempt people to try to find a 'side door' way of avoiding the Transformed Way Of Living.

Finding a 'side door' way of avoiding taking up the Transformed State
and maintaining your obsoleted artificial ways of validating yourself

[47] To describe the most common example of this practice of 'Side Dooring'. It is natural and beneficial to continue studying the explanations of the human condition and deepening your understanding of yourself and the world, but when the cornered Mexican Standoff feelings develop, this guzzling of the information can be developed into a way of avoiding those cornered feelings. Certainly everyone needs to be using these understandings to help make sense of their lives and the world around them, and even to heal their version of humanity's 2 million years of psychological upset, but to

focus on doing that while avoiding adopting the Transformed Way Of Living when that is so needed to solve the world's problems is, at base, another way of selfishly hanging onto, rather than letting go, the now obsoleted egocentric, 'I'm a legend, castle building' way of living.

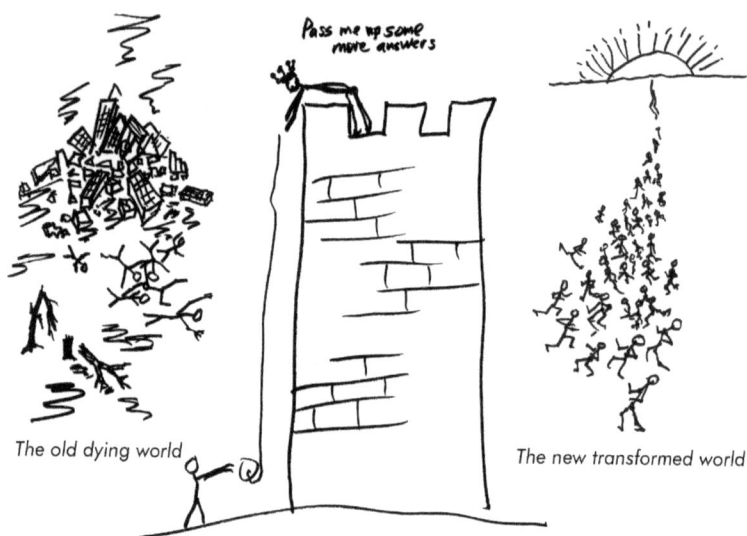

The old dying world

Pass me up some more answers

The new transformed world

Drawing by Jeremy Griffith © 1996-2023 Fedmex Pty Ltd

The crime of guzzling the information to grow your ego castle while avoiding the Transformed Way Of Living that's needed to save the world

[48] Making sense of your existence and healing your psychosis are very important activities to pursue, but they have to be tempered by the priority of the need to take up the Transformed Way Of Living that saves the world. And it has to be remembered that the Transformed Way Of Living is the best form of therapy for everyone's corrupted condition, so taking it up is actually the basis of the therapy that all humans need.

[49] These understandings do finally make the real therapy of our corrupted lives possible—and I and others in the Sydney WTM are writing a book about that real therapy that is now possible [you can

read *Therapy For The Human Condition* on our website], so we fully recognise the importance of therapy—but while the priority is to save the world from imminent self-destruction through the adoption of the Transformed Way Of Living, such a selfish focus is another obscenity.

[50] There is a similarity between the Transformed Way Of Living and religious transformations that we can learn from. As I emphasise in chapter 9:5 of *FREEDOM*, the Transformed Way Of Living is *not* a religion, but both ways of living free you from your corrupted condition by living in support of something pure; in the case of religion, the pure life of a prophet, and in the Transformed Way Of Living, of a human-condition-understood, pure new world. Well, in the case of Christianity, the 12 disciple first followers of Christ didn't properly see how the faith that Christ was introducing worked. Venerating and worshipping Christ as they did was not the point. The real point of Christianity is to let go your embattled, overly-upset, self-preoccupied way of living and instead defer to and live through your support of Christ. However, just as people have resisted letting go their upset way of living and taking up the Transformed State where you live in support of the redeeming understanding of ourselves, so the disciples also resisted letting go their upset way of living and deferring to and living in support of Christ, complaining that **'This is a hard teaching. Who can accept it?'** (John 6:60). It was actually St Paul, who came along afterwards, who realised the immense 'fall off your donkey' relief that living in support of Christ rather than in the way your corrupted condition wanted you to live, was how Christianity works. The disciples loved the soundness of Christ, just as people love the sound understandings of the human condition, but it is living in support of the understanding of the human condition that transforms your life and saves humankind. Unless you take up the Transformed Way Of Living you are not making proper use of being able to understand the human condition.

[51] The truth is that the various 'pocketing the win' strategies are stupid ways to live given the world-saving importance and phenomenal excitement and satisfaction of taking up the Transformed Way Of Living. So 'pocketing the win' is not an attitude that will be popular for long.

[52] As I mention in paragraph 1209 of *FREEDOM*, there are other ways of avoiding the subconscious Mexican Standoff feelings of being exposed and cornered by these understandings that dictate you leave behind your power, fame, fortune and glory 'castle building', or your pseudo idealistic ways of feeling good about yourself, and take up the Transformed Way Of Living. These include finding supposed faults with myself or the WTM, such as saying "Since Jeremy isn't resigned he can't properly appreciate the lives of resigned people, so I KNOW BETTER, and this is what everyone should do, follow me". There is the strategy of 'SIDING WITH THE ANGELS' as we call it, where you delude yourself you are an innocent person uncorrupted by the human condition and so don't need to become transformed. Then there is 'SHIPS AT SEA' ARROGANCE where people who have managed to largely avoid resigning to living in denial of the human condition during their adolescence—those who refused to 'pull into a port' to escape the 'storms out at sea' that confronting the human condition causes—feel so righteous and even innocent for not having resigned that they are not humble enough to admit their insecure egocentricity and resulting need to be transformed (see F. Essay 60 for a description of 'Ship at Sea Arrogance'). Other avoidance strategies include saying "THESE IDEAS AREN'T ORIGINAL anyway"; and "I HAD ALREADY COME UP WITH THESE IDEAS"; and, for supporters of the new age, politically correct, 'woke' movements, deludedly saying "It's NICE TO HAVE YOU JOIN US in creating a compassionate world".

The 'I know better' deluded strategy

'Siding with the angels' and 'Ships at sea arrogance' strategies where you delude yourself you are an innocent person

The 'I had already come up with these ideas' defensive strategy

The 'These ideas aren't original' defensive strategy

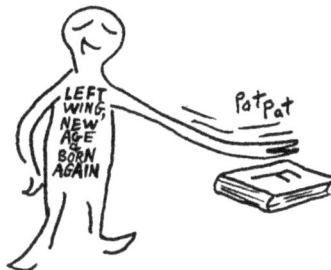

The 'It's nice to have you join us in creating a compassionate world' deluded strategy

[53]Clearly what is needed in this early 'shock of change' stage where all manner of procrastinations, exploitations and avoidances inevitably occur is for exceptionally gifted, clear-sighted, St-Paul-like individuals to appear who are able to LEAD THE WAY TO A WORLD WHERE EVERYONE IS ENJOYING THE ALL-MEANINGFUL AND ALL-EXCITING TRANSFORMED WAY OF LIVING—which is illustrated in the drawing of *Humanity's Journey from Ignorance to Enlightenment* by the person excitedly running toward the all-wonderful transformed world.

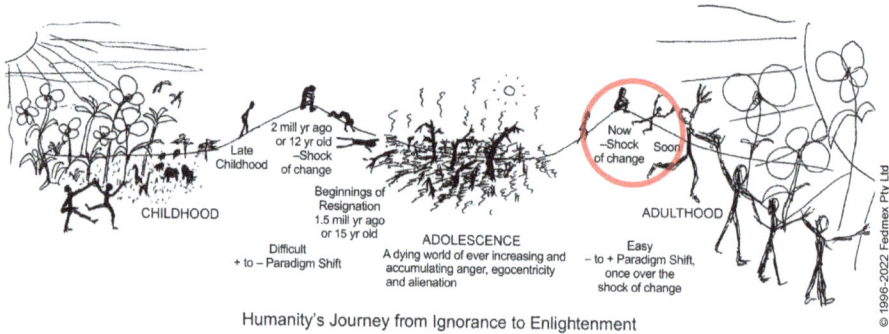

Humanity's Journey from Ignorance to Enlightenment

[54]WTM founding member Anthony [Tony] Gowing, with many other WTM founding members and WTM Centre members close behind him, is just such a leader of the great break-out for the human race from the fast looming threat of the terminal alienation and extinction of our species into a warm sun/understanding-filled Transformed State of unimaginable relief and happiness! You can read and hear Tony's inspired teachings about the Transformed Way Of Living [along with everything else you need to know about the Transformed State] on the Transformation page on our website.

[55] So finally, humanity is coming back home to peace, happiness, soundness and sanity.

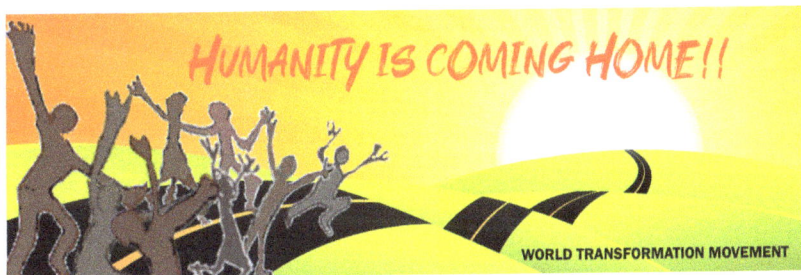

- - - - - - - - - - - - - - - - - -

Addendum – A summary of the human journey using caricatures

[56] I thought it might be helpful to conclude this presentation with a summary of the human journey using caricatures I've drawn over the years of all the different stages in our species' and in our own immensely heroic journey.

[57] All these stages are described in chapter 8 of *FREEDOM*.

[58]This first drawing of a mother and child symbolises the nurturing, love-indoctrination process that created our cooperative and loving moral soul and formed the original fully integrated state our ancestors once lived in, as depicted in the second drawing of the three people hugging each other.

[59] The next drawing illustrates how, when we became conscious some 2 million years ago, our original cooperative and loving instinctive self, the 'voice' or expression of which is our conscience, in effect criticised our self-managing conscious mind's necessary search for knowledge, which upset us, causing us to become defensively angry, egocentric and alienated—sufferers of the guilt-stricken, depressed state of the human condition.

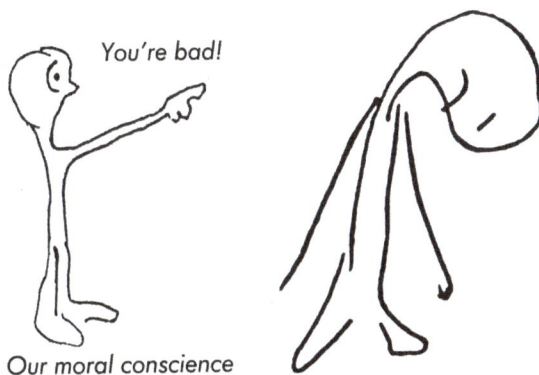

You're bad!

Our moral conscience

J.Griffith © 1990 Fedmex Pty Ltd

[60] This next drawing shows us trying to understand *why* we corrupted our soul and became sufferers of the agony of the human condition.

J.Griffith © 1996 Fedmex Pty Ltd

[61]Then, unable to understand why we had 'fallen from grace', in other words become corrupted, we eventually, from about 1.5 million years onwards, or at about 15 years of age in the case of our individual lives, began to resign to having to live in denial of our corrupted condition and take up a life of seeking the artificial reinforcements of power, fame, fortune and glory to sustain our sense of self-worth.

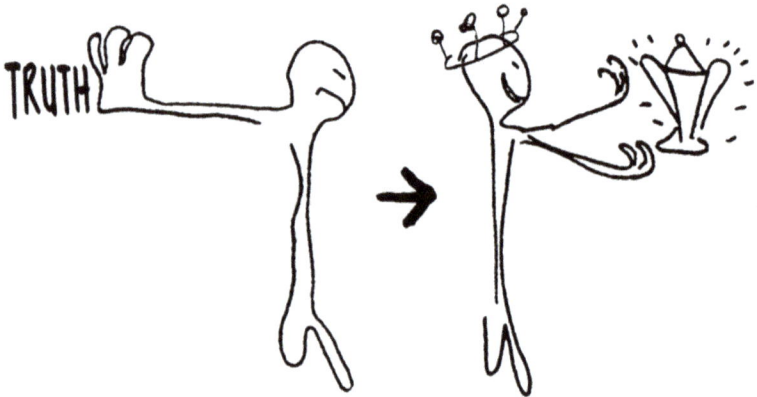

J. Griffith © 1990 Fedmex Pty Ltd

[62] These next two drawings show how this defensive angry, egocentric and alienated way of living developed more and more as our heroic search for knowledge—ultimately for self-knowledge, understanding of our corrupted human condition—continued. The first image describes how materialism, like acquiring fast cars and so forth, serviced our need for escapism and ego reinforcement.

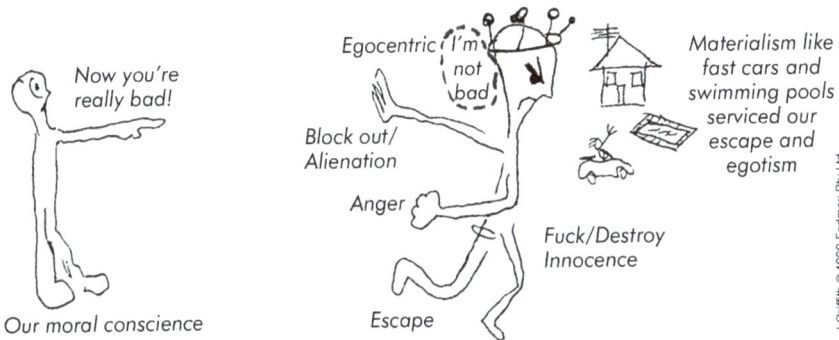

Now you're really bad!

Our moral conscience

Egocentric *I'm not bad*

Block out/ Alienation

Anger

Escape

Fuck/Destroy Innocence

Materialism like fast cars and swimming pools serviced our escape and egotism

J.Griffith © 1990 Fedmex Pty Ltd

[63] The second image shows upset developing to the extent that we become *extremely* egocentric and in need of artificial reinforcement, and with fierce defensive responses to any perceived criticism.

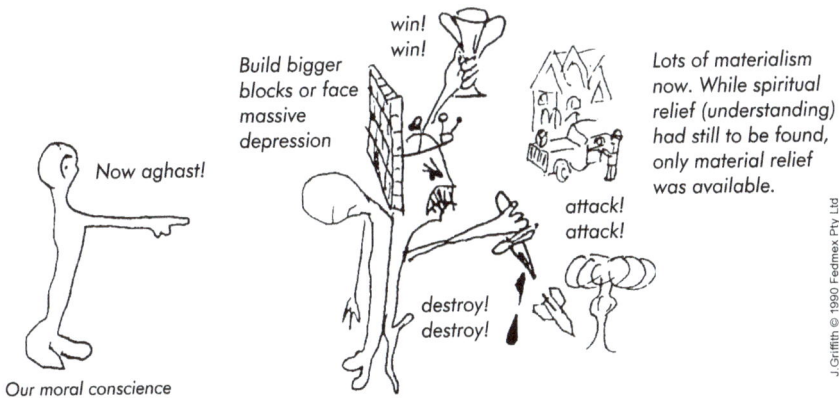

Now aghast!

Our moral conscience

Build bigger blocks or face massive depression

win! win!

destroy! destroy!

attack! attack!

Lots of materialism now. While spiritual relief (understanding) had still to be found, only material relief was available.

J.Griffith © 1990 Fedmex Pty Ltd

[64]This next drawing shows how upset eventually became so extreme that we adopted all manner of artificial, pseudo idealistic, feel-good, virtue-signalling ways of escaping feeling that we were bad and unworthy.

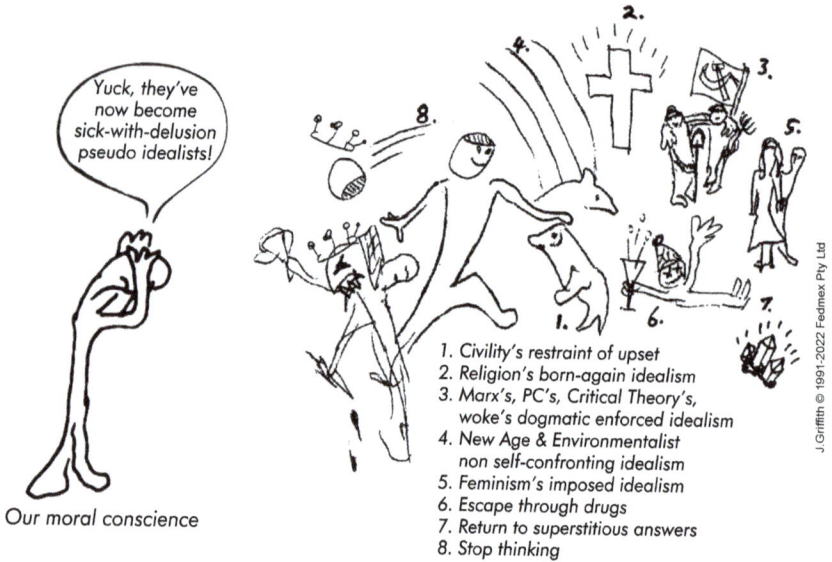

Yuck, they've now become sick-with-delusion pseudo idealists!

Our moral conscience

1. Civility's restraint of upset
2. Religion's born-again idealism
3. Marx's, PC's, Critical Theory's, woke's dogmatic enforced idealism
4. New Age & Environmentalist non self-confronting idealism
5. Feminism's imposed idealism
6. Escape through drugs
7. Return to superstitious answers
8. Stop thinking

J.Griffith © 1991-2022 Fedmex Pty Ltd

[65] Eventually, disillusionment with the dishonest, deluded pseudo idealistic life forced many to return to the earlier soul-corrupting-but-knowledge-finding, power-fame-fortune-and-glory-hunting to sustain your self-esteem, which only led to even greater levels of upset, which is the hollow, no-soul-left-at-all, extreme state of upset.

Terminally alienated hero of the story of life on Earth

Our moral conscience

J. Griffith © 2011 Fedmex Pty Ltd

[66] The overall situation the human race had then arrived at was the completely alienated and mad state it has now ended up in.

J. Griffith © 1996 Fedmex Pty Ltd

[67] But *then*, on the threshold of terminal alienation and the extinction of our species, humanity's heroic efforts to accumulate knowledge has finally led to the finding of the reconciling, redeeming and healing understanding of our corrupted condition, which obsoletes all our old artificial power, fame, fortune and glory ways of sustaining our sense of self-worth.

J.Griffith © 1990 Fedmex Pty Ltd

[68] The initial problem the arrival of this dreamed-of understanding faces is that we've been living in such fear and denial of our corrupted human condition that we struggle to take in or 'hear' the redeeming and liberating explanation of it. We suffer from the Deaf Effect.

J.Griffith © 1996 Fedmex Pty Ltd

[69] But then, after persevering in reading and listening to all the understandings of the human condition, we overcome the Deaf Effect and gain access to these utterly relieving and exciting understandings.

[70] But then, as we 'guzzle' these understandings that finally make sense of ourselves and the world, we can become confronted by the exposure of our corrupted condition that the understanding causes. Basically, we can't absorb and adjust to all the liberating understandings fast enough to not experience some remaining fearful exposure from all the truth about how corrupted we actually are. In other words, while this 'exposure day' or 'judgment day' is actually a 'day' of compassionate understanding, not a 'day' of condemning judgment, there is still a potential to feel a degree of difficult self-confrontation.

[71] The solution to cope with that exposure is to not overly study the information more than our degree of soundness of self can tolerate. We only need to investigate the information sufficiently to verify that it's true, and then take up the Transformed Way Of Living where we focus on living in support of this information that's needed to save the world. If we study the information more than our degree of soundness of self can cope with we risk becoming furiously angry and defensive towards the information, an unconscionable response because what we would be doing is condemning the human race to terminal alienation and extinction. Everyone should verify the truth then support it at a comfortable distance from it.

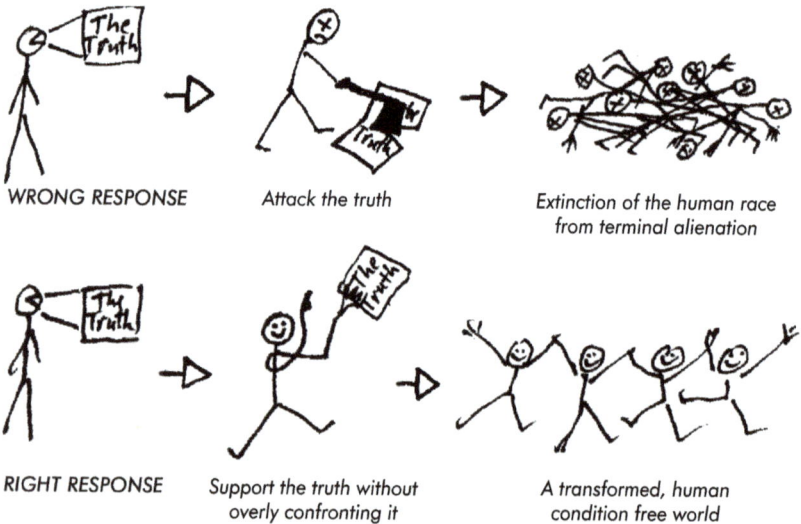

WRONG RESPONSE   Attack the truth   Extinction of the human race from terminal alienation

RIGHT RESPONSE   Support the truth without overly confronting it   A transformed, human condition free world

Drawing by Jeremy Griffith © 1996-2014 Fedmex Pty Ltd

[72] The next problem is the 'Mexican Standoff' where we don't want to deny, or, worse still, attack the truth, but at the same time we are so habituated to living off our 'power and glory' artificial ways of sustaining our sense of self-worth that we resist taking up the Transformed Way Of Living. We feel stuck, unable to go back and unable to go forward!

J. Griffith © 2019 Fedmex Pty Ltd

[73] In fact, this 'Mexican Standoff' can tempt us to 'pocket the win' as we call it—focus on benefiting from being able to understand ourselves and the world rather than focus on taking up the Transformed State where we live in support of these understandings that alone can save the world. We can even be tempted to use the understanding to make ourselves an even more successful 'power and glory', egocentric 'castle-builder', the opposite of letting that way of living go and becoming transformed.

[74]Eventually, however, everyone realises that the Transformed Way Of Living is the only responsible way to live, and it becomes a universal practice.

[75] And the human race is then freed at last from the human condition, and joy and happiness comes to planet Earth!

[76] Thank you for watching this presentation.